Mobil New Zealand Nature Series

Common Ferns and Fern Allies

Mobil New Zealand Nature Series

Common Ferns and Fern Allies

Brian Parkinson

REED

Introduction

In no other country in the world are ferns so closely bound up with a nation's pysche as they are here in New Zealand. Our national sporting teams, exported products, and even our national airline are adorned with fern fronds. Think of our wild, green places and you are almost certain to think of ferns flourishing in verdant splendour.

In fact New Zealand has only about 200 of the approximately 10,000 species now known. Despite this, ferns have evolved here into a multitude of shapes and forms designed to cope with just about every environment, from tussock grasslands to geothermal areas, storm-battered coastal cliffs to steep limestone cliffs, and everything in between.

Ferns and their allies are an extremely ancient family of plants. They first evolved in the mid-Palaeozoic era, and by the Carboniferous period of some 350 million years ago were among the dominant plant groups on the planet, both in numbers and in size. Huge tree ferns and large tree-like club mosses flourished and, except for their size, would not have looked very different from those ferns around us today.

Ferns and their allies form a natural group of plants that is classified by botanists into the following classes: the Filicopsida, which are the true ferns, and the various fern allies — the Psilotopsida (Psilotaceae) and the Equisetopsida (horsetails). Another class that is often included is the Lycopsida, which comprises clubmosses, spikemosses and quillworts, although it is probable that these are only distantly related to ferns.

The easiest way to differentiate between the ferns and the fern allies is by looking at their leaf structures. Those of the ferns are called megaphylls. They are large and are the fern fronds that we are all familiar with. Those of the fern allies are small structures called microphylls, some of which are scale-like.

This book covers 65 species and includes the most common native ferns and fern allies that an interested observer is likely to encounter in New Zealand. In addition, several exotic species are also featured as they are now some of our commonest ferns, particularly in the more northern parts of the country.

Except for the tree ferns and the king fern, all the ferns discussed in this book are medium-sized plants, growing to about 50–100 cm.

Fern taxonomy

A source of wonder to many lay people is why, when they are looking in a book for a common fern such as bracken, they find it is listed under the incomprehensible and probably unpronounceable name of *Pteridium esculentum*.

Scientific names are not given, as has been suggested, just so that scientists can feel superior. They are used because they are universal and make perfectly clear just what that particular species of flora or fauna is. For example, bracken is found in a great many countries and probably has several hundred different common names depending upon where it occurs.

Scientific names are made up of two parts. The first name *Pteridium* tells you which genus or group bracken belongs to. It is roughly equivalent to a surname, so that you know what its closest relatives are. The second name is that of the species, here *esculentum*, and tells you which particular individual fern is being discussed.

These names are usually Latin (as this, being a 'dead' language, does not change) and should ideally describe a characteristic of the species being discussed. *Pteridium* means 'fern-like' and *esculentum* translates as 'edible'.

Medicinal uses of New Zealand plants

Many New Zealand plants, including ferns, were used by the Maori and also by European settlers for medicinal uses. Where possible, this information has been included for its historical interest. However, readers should note that this book is not intended as a practical guide to the medicinal use of New Zealand plants, and should not be used as such.

Parts of a fern

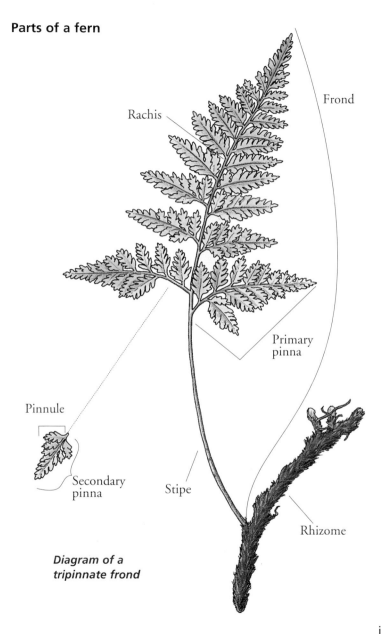

Frond

Rachis

Primary
pinna

Pinnule

Secondary
pinna

Stipe

Rhizome

*Diagram of a
tripinnate frond*

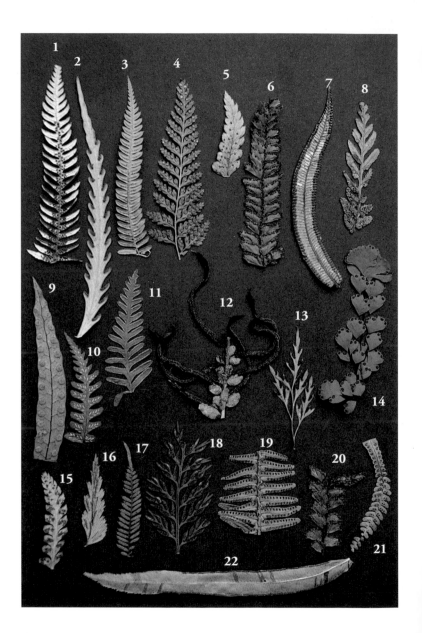

Fern spores

The patterns of fern sori, or spore-bearing parts, vary greatly between genera, and fern taxonomists often use these patterns to help in classification. The plate opposite illustrates some of the many different patterns of positions of sori that occur.

1	*Cyathea dealbata*	12	*Blechnum novae-zelandiae*
2	*Asplenium flaccidum*	13	*Asplenium terrestre*
3	*Cyathea medullaris*	14	*Adiantum raddianum*
4	*Lastreopsis glabella*	15	*Polystichum richardii*
5	*Polystichum richardii*	16	*Asplenium polyodon*
6	*Adiantum hispidulum*	17	*Blechnum fraseri*
7	*Marattia salicina*	18	*Asplenium terrestre*
8	*Todea barbara*	19	*Nephrolepis cordifolia*
9	*Microsorum postulatum*	20	*Adiantum hispidulum*
10	*Deparia petersenii*	21	*Nephrolepis cordifolia*
11	*Pteris tremula*	22	*Pteris cretica*

Native tree ferns

1 *Cyathea medullaris*
Black tree fern / Mamaku

Family CYATHEACEAE

Description
- Trunk black; fibrous at base; grows to 20 m high; scarred by fallen fronds.
- Fronds thick, long and horizontal; midrib turns black in mature specimens.

Distribution & Habitat
- Common in damper areas of the North and South Islands, but rare and local on Stewart Island.

- Also found on several Pacific islands from Fiji to Pitcairn.
- Favours damp, shady gullies, but also relatively common in pine forests.

Notes
- Edible. The Maori baked the pith of mamaku in earth ovens.
- Emergent fronds are very susceptible to possum damage.

◀ *Cyathea medullaris*

Cyathea dealbata
Silver tree fern / Ponga

Family CYATHEACEAE

Description
- Trunks up to 10 m in height, covered in light brown, peg-like stipe bases.
- Fronds are long and horizontal with distinctive pale, whitish undersides.

Distribution & Habitat
- In damp bush from lowlands to moderate altitudes from North Cape to Dunedin, but largely absent from Westland, Fiordland and Stewart Island.

Notes
- The pale underside of fronds were used as markers by Maori war parties.
- Pakeha settlers used ponga trunks to build the walls of their huts.
- Hybridises readily with the Australian species *C. cooperi*.

◀ *Cyathea dealbata*

3 *Cyathea smithii*
Soft tree fern / Katote

Family CYATHEACEAE

Description
- Trunk brown; to 6 m tall; often noticeably wider at base owing to matted fibrous rootlets.
- Fronds long, horizontal, light green and soft to touch.
- Grass-like skirt of dead stipes.

Distribution & Habitat
- An endemic species confined to colder, damper, montane areas in the north, but spreading to lowlands in the South and Stewart Islands.

Notes
- Edible. The heart of the katote was cooked in earth ovens by the Maori.
- Katote secretes a substance inhibiting the root growth of competing plants.

◀ *Cyathea smithii*

Dicksonia squarrosa
Rough tree fern / Wheki

Family DICKSONIACEAE

Description
- Sometimes has multiple trunks.
- Trunks are covered with rough stipe bases; grow to 7 m.
- Fronds often form a scruffy crown with an untidy skirt of dead fronds.

Distribution & Habitat
- Abundant sub-canopy species from the Three Kings Islands in the north to the Chatham Islands in the south. Often seen regenerating in burnt-off areas.

Notes
- Traditionally used by the Maori to reduce bleeding.
- Often grown to control erosion and stabilise road cuttings.
- Maori traditionally used wheki to build whare walls and pa palisades (tuwatawata).
- On Stewart Island, this fern suffers from deer browsing.

◀ *Dicksonia squarrosa*

5 *Dicksonia fibrosa*
Fibrous tree fern / Wheki-ponga

Family DICKSONIACEAE

Description
- Dense crown of short, harsh-textured, narrow fronds.
- Trunk of soft, fibrous, compressed matted rootlets, often massive in older specimens.
- Usually grows as a single trunk, unlike *Dicksonia squarrosa*.
- Heavy skirt of dead fronds.

Distribution & Habitat
- Widespread throughout all three main islands and the Chathams, but only really common in mountainous areas of the central North Island.
- Can tolerate more open sites than other tree ferns.

Notes
- Traditionally used by Maori and early settlers for making rat-proof storage containers.
- The slowest–growing of all our tree ferns, the biggest specimens are often several hundred years old.

◀ *Dicksonia fibrosa*

▲ *Adiantum aethiopicum*

▼ *Adiantum cunninghamii*

Native ground ferns

6 *Adiantum aethiopicum*
True maidenhair / Tawatawa

Family ADIANTACEAE

Description
- Rhizomes creeping.
- Stipe smooth and brownish black.
- Leaf segments thin and pale green, heart- or fan-shaped, and sometimes deeply notched.

Distribution & Habitat
- Common in the North Island from North Cape to Waikato, but more local in southern areas.
- Also found in Australia, New Caledonia and Africa.
- Grows in lowland and coastal areas in scrub and light forest. Sometimes found on sheltered banks.

Notes
- Very common in cultivation.
- Named for Ethiopia, but not found in Africa.
- Ancient Romans and Greeks regarded this fern as a hair restorer.

7 *Adiantum cunninghamii*
Common maidenhair / Tawatawa

Family ADIANTACEAE

Description
- Rhizomes creeping, often forming patches.
- Fronds bright green with a bluish flush.
- Stipe black or blackish-brown and shining, hairy at base.

Distribution & Habitat
- From the Kermadec Islands to the Chathams, and the only maidenhair to be found on Stewart Island.
- Locally common in coastal areas, under short scrub and on creek banks.

Notes
- An endemic species.
- Our most widespread maidenhair.

▲ *Adiantum diaphanum*

▼*Adiantum hispidulum*

8 *Adiantum diaphanum*
Small maidenhair / Tawatawa

Family ADIANTACEAE

Description
- Rhizomes short and erect.
- Stipes and rachises smooth and brownish.
- Fronds dark green with small, delicate, oval leaf segments.

Distribution & Habitat
- Common from North Cape to Waikato but found only locally from there south to Nelson.
- Also occurs in Australia and on some Pacific islands north to China and Japan.
- Coastal and lowland forest, often growing in permanently damp situations near water-falls and water seepages.

Notes
- Seemingly delicate but surprisingly hardy in cultivation.

9 *Adiantum hispidulum*
Rosy maidenhair / Tawatawa

Family ADIANTACEAE

Description
- Newly emergent fronds bright rosy pink.
- Short-creeping, branching rhizomes covered with brown scales.
- Leaf segments green, covered with short white hairs.

Distribution & Habitat
- Common from North Cape to Bay of Plenty, but from there found locally south to Wairarapa.
- Also found from Polynesia through Southeast Asia to India.
- Grows on dry, sunny banks and in rocky areas; also under light scrub.

Notes
- Our only maidenhair with pink-coloured emergent fronds.
- One of the easiest maidenhairs to cultivate.

▲ *Anarthropteris lanceolata*

▼ *Asplenium bulbiferum*

10   *Anarthropteris lanceolata*
Lance fern / Ngarara whare

Family GRAMMITIDACEAE

Description
- Sometimes occurs in dense patches.
- Fronds undivided and slightly fleshy, light green above but paler below.
- Large sori on lower surface of fronds cause prominent bumps on upper surface.

Distribution & Habitat
- Common in North Island, and to about Banks Peninsula and Punakaiki in the South Island, in lowland and coastal forest.
- Occurs epiphytically on trees; also on banks and large rocks.

Notes
- One of our few endemic genera.

11   *Asplenium bulbiferum*
Hen and chickens fern / Mouku

Family ASPLENIACEAE

Description
- Rhizomes short, erect and scale-covered.
- Stipes and rachises green above, brown below.
- Fronds graceful and arching and often covered with small fernlets (bulbils), hence its common name.

Distribution & Habitat
- Widespread in lowland and low montane forests throughout the three main islands, except in drier inland areas of Otago.
- Also occurs in Australia and Tasmania.

Notes
- There is a closely related species, *A. gracillimum*, but this lacks bulbils.
- Newly emergent fronds were eaten by Maori.
- Now much reduced in many areas owing to deer and possum browsing.
- Popular in the nursery trade in many countries.

Asplenium flaccidum
Hanging spleenwort /
Nga makawe o raukatauri

Family ASPLENIACEAE

Description
- Very common but highly variable species.
- Dark matt green in colour, with somewhat leathery fronds.
- Epiphytic form produces long hanging fronds, dark green and leathery.
- Terrestrial forms have shorter, stouter fronds.

Distribution & Habitat
- Widely distributed in lowland forests of main and offshore islands.
- Also grows in Australia.
- Also occasionally found in rocky areas and in pine plantations.

Notes
- The Maori name refers to hanging ringlets, which the fronds resemble.

◀ *Asplenium flaccidum*

▲ *Asplenium oblongifolium*

▼ *Asplenium obtusatum*

13 *Asplenium oblongifolium*
Shining spleenwort / Huruhuruwhenua

Family ASPLENIACEAE

Description
- Very variable fern, the shape of which largely depends on habitat.
- Leaf segments shining green above, matt green below. Oblong in shape with finely-toothed margins.
- Sori large and conspicuous.

Distribution & Habitat
- From the Kermadec Islands to the Chathams, but absent from lower South Island and Stewart Island.

- A fairly coarse fern, which forms large clumps in ideal conditions.
- Often epiphytic, but more frequently terrestrial near the coast and in pine plantations.

Notes
- The Maori name translates as 'entirely shining'.
- A hardy fern which is common in cultivation but susceptible to slug damage.

14 *Asplenium obtusatum*
Shore spleenwort / Paranako

Family ASPLENIACEAE

Description
- Rhizomes thick and woody.
- Fronds very thick and fleshy, stiffly erect, yellowish-green to darker green in colour, some-times occuring in thick clumps.

Distribution & Habitat
- Throughout country from Auckland south to the Chathams and the sub-Antarctic islands, but rarer on the east coast.

- Widespread in South Pacific from New Zealand to Easter Island; also in South America.
- Common in coastal areas on rocks and banks and in scrub and forest, but rarer inland.

Notes
- Replaced in Northland by the similar fern *Asplenium northlandicum*.

▲ *Asplenium polyodon*   ▼ *Azolla filiculoides* var. *rubra*

15 *Asplenium polyodon*
Sickle spleenwort / Petako

Family ASPLENIACEAE

Description
- Rhizomes short and creeping.
- Fronds dark glossy green but paler below with heavily serrated margins.
- Sori in radiating lines.

Distribution & Habitat
- Common throughout country from the Kermadec Islands to the Chathams.
- Also found in Australia, New Guinea, Southeast Asia, Madagascar and Pacific Islands.
- In coastal lowland to moderate altitude forests.
- Usually epiphytic, but sometimes on the ground growing in large, arching fronds.
- Not uncommon in pine forests.

Notes
- Slow-growing but hardy.
- Sometimes also called Shark-tooth fern.

16 *Azolla filiculoides* var. *rubra*
Red azolla / Retoreto

Family SALVINIACEAE

Description
- A free–floating, aquatic fern that forms colonies on still waters.
- Generally propagates vegetatively, but fertile plants occasionally occur.
- Also occurs in Australia, Africa and the Americas.

Notes
- Forms dense red 'carpets' on still waters.

Distribution & Habitat
- Common in warmer areas of the North Island.

Blechnum colensoi
Colenso's hard fern / Peretao

Family BLECHNACEAE

Description
- Very distinctive species.
- Sterile fronds up to 110 cm long; dark green, leathery and drooping, varying from strap-shaped to deeply lobed.
- Fertile fronds are also long, but are distinctively black and narrow.

Distribution & Habitat
- In damp lowland and montane forests south of a line from Thames to Port Waikato.
- Favours damper areas along banks of heavily wooded, flowing streams and in the spray zone near waterfalls.
- Can tolerate permanently damp fronds.

Notes
- An endemic species but with closely related species in Asia.
- Formerly called *Blechnum patersonii*, which is an Australian species.

◀ *Blechnum colensoi*

Blechnum discolor
Crown fern / Piupiu

Family BLECHNACEAE

Description
- Rhizomes are erect and some-times form short, woody trunks.
- Sterile fronds erect in mature plants, forming a crown.
- New fronds pale green, paler below, and darkening on the upperside as plants mature.
- Fertile fronds erect, black, 'fishbone'-shaped, appearing in winter.

Distribution & Habitat
- Common throughout the coun-try in drier lowland forest, but can reach montane areas in the North Island.
- Often forms extensive colonies in open beech forest and scrub.

Notes
- An endemic species that is rea-sonably common in cultivation.
- In Maori mythology, kakapo held fronds of piupiu above their heads to avoid detection.

◀ *Blechnum discolor*

Blechnum filiforme
Thread fern / Panako

Family BLECHNACEAE

Description
- Long, creeping rhizomes are covered with short, sterile fronds at ground level.
- Becomes epiphytic when a tree is encountered, growing up the trunk and producing long pendulous fronds.
- Fertile fronds are always epiphytic and bear long, thread-like pinnae.

Distribution & Habitat
- Common in lowland areas in the North Island and in the Nelson area of the South Island, particularly in areas of puriri, nikau and tree ferns.

Notes
- A remarkable fern which can cover large areas of ground, unique among our ferns because of its three different frond shapes.

◀ *Blechnum filiforme*

▲ *Blechnum fluviatile*

▼ *Blechnum fraseri*

20 *Blechnum fluviatile*
Kiwikiwi

Family BLECHNACEAE

Description
- Rhizomes short and stout.
- Sterile fronds long and greenish-brown, often touching the ground.
- Fertile fronds dark and erect with stiff, upward-pointing pinnae.

Distribution & Habitat
- Common throughout the country, particularly in montane areas of the North Island along stream and river banks, where it often forms large colonies.

Notes
- Forms attractive, spreading rosettes.
- Also found in Tasmania and southeastern Australia.
- Contains a powerful natural insecticide.

21 *Blechnum fraseri*
Hard fern

Family BLECHNACEAE

Description
- Fern has thin, spindly trunks, distinguishing it from other *Blechnum* species.
- Mature fronds dark green and glossy, contrasting with the paler fronds emerging from centre of plant.
- Fertile fronds are similar to sterile fronds, which is not typical of *Blechnum* species.
- The only New Zealand *Blechnum* with bipinnate fronds (i.e., the primary pinnae bear smaller pinnules).

Distribution & Habitat
- In kauri and other lowland forest from North Cape to Tauranga and the King Country.
- Also in coastal areas of Westland and northwest Nelson. Prefers drier, well-drained areas.
- Also found in Indonesia and the Philippines.

Notes
- Differs from other *Blechnum* species and probably needs further taxonomic revision.

Blechnum novae-zelandiae
Cape fern / Kiokio

Family BLECHNACEAE

Description
- Sterile fronds in damp and shady conditions are long, green and pendulous.
- Fronds are red when young.

Distribution & Habitat
- Common throughout country from the Kermadec Islands to the Chathams, being absent only from South Island montane areas.
- Probably the most widely distributed of our ferns, being particularly prolific in disturbed areas such as roadside cuttings.

Notes
- For a long time *B. novae-zelandiae* was confused with *Blechnum capense*, which is a South African species named for the Cape Province. It is only recently that this fern has been given its own scientific name, but it will probably continue to be known as *B. capense* by most people.
- The emergent fronds, called 'fiddlesticks', of the kiokio were a popular food item among the Maori.

◀ *Blechnum novae-zelandiae*

▲ *Blechnum pennamarina*   ▼ *Blechnum procerum*

23 *Blechnum pennamarina*
Alpine hardfern

Family BLECHNACEAE

Description
- Sterile fronds pink, turning dark green as plant matures; relatively short and coarse to touch.
- Fertile fronds longer, dark brown, and 'fishbone'-shaped.
- Rhizomes are creeping.

Distribution & Habitat
- Common in South Island and southern North Island, but local from Bay of Plenty north.

- Found also on the Chathams and the sub-Antarctic islands.
- Grows in open grassland and along bush margins; common in alpine herbfields.
- Also in Australia and South America.

Notes
- Sometimes called Alpine water fern.
- A common fern in cultivation.

24 *Blechnum procerum*
Small kiokio

Family BLECHNACEAE

Description
- Fronds bronzy-green, coarse and thick, with the shortest pinnae at the bottom of the frond and the longest pinna at the top.
- Fertile fronds dark and thin, rising from centre of plant.

Distribution & Habitat
- Common throughout country

at moderate altitudes, and down to sea level in south.
- Usually clumped, but sometimes forms thick patches in drier, more open forests.

Notes
- Readily hybridises with *B. novae-zelandiae* where the two plants meet.

▲ *Deparia petersenii*

▼ *Doodia media*

25 *Deparia petersenii*
Deparia

Family DRYOPTERIDACEAE

Description
- Rhizomes long-creeping, fleshy and scaly.
- Fronds erect, light green above and paler below with grooved stipes.

Distribution & Habitat
- From North Cape to about central Westland, but extending its range southwards.
- Also in Australia, Asia and on many Pacific Islands.
- Favours bush margins, shady stream banks and scrubby areas.

Notes
- Sometimes erroneously known as *Lunathyrium japonicum*, but this species does not occur in New Zealand.

26 *Doodia media*
Rasp fern / Pukupuku

Family BLECHNACEAE

Description
- Rhizomes short-creeping, sometimes erect.
- Newly emergent fronds pink; mature fronds green, coarse in texture and harsh to touch.
- Sori abundant along centre of pinnae.

Distribution & Habitat
- Grows in fairly exposed areas and is common in secondary growth, especially in coastal areas and under pohutukawa.
- Distributed from North Cape south to Nelson and the Marlborough Sounds.
- Also occurs on Norfolk Island and in Australia from Tasmania north to Queensland.

Notes
- Pukupuku is Maori for 'goose-flesh', a reference to this fern's texture.
- An easily cultivated species.

▲ *Doodia mollis*

▼ *Gleichenia dicarpa*

27 *Doodia mollis*
Small rasp fern / Mokimoki

Family BLECHNACEAE

Description
- Rhizomes erect.
- Sterile fronds erect and pink when young, hard and coarse to touch.
- Fertile fronds prostrate near to ground.

Distribution & Habitat
- Usually in coastal forest or open scrub.

- Locally abundant from North Cape to Hawke's Bay.

Notes
- Can be confused with *D. squarrosa*, but fronds are more sickle-shaped and sparsely situated along rachis.

28 *Gleichenia dicarpa*
Swamp umbrella fern / Waewaekaka

Family GLEICHENIACEAE

Description
- Rhizomes long-creeping, multi-branched and wiry.
- Stipes red-brown and wiry.
- Fronds variable, dark green, up to 100 cm long, with several tiers of opposite pinnae.
- Ultimate segments small and pouched.

Distribution & Habitat
- Common throughout country in open areas and swamps, but at higher altitudes in the south. Sometimes found in thermal areas.

- Also occurs in Tasmania and eastern Australia.
- Often forms dense, tangled thickets with pinnae interlacing, sometimes growing up through scrub.

Notes
- Can be confused with *G. microphylla*.
- The Maori name translates as 'footprint of the kaka'.

▲ *Histiopteris incisa*   ▼ *Hymenophyllum demissum*

29 Histiopteris incisa
Water fern / Mata

Family DENNSTAEDTIACEAE

Description
- Rhizomes thick, creeping and often covering wide areas.
- Fronds soft, highly-divided, upright, pale bluish-green to dark green in colour.
- Often browns off in winter.

Distribution & Habitat
- Common throughout country from the Kermadecs to the sub-Antarctic islands.
- Also in southern Africa, the Americas, Australia and Southeast Asia.
- Favours clearings, bush margins and disturbed ground such as roadside cuttings.
- Also in thermal areas.

Notes
- Evergreen in the more northern areas, but deciduous in the south.
- In Australia known as the Bat's wing fern.

30 Hymenophyllum demissum
Filmy fern / Irirangi

Family HYMENOPHYLLACEAE

Description
- Rhizomes wiry and creeping.
- Fronds stiffly arching up to 25 cm long, pale glaucous green with smooth-edged, blunt-tipped pinnae.
- Sori usually arranged in pairs.

Distribution & Habitat
- Common throughout country from the Kermadecs to the Chathams and the Auckland Islands, except for drier parts of Canterbury and Otago.
- In lowland and moderate altitude montane forest.
- Usually terrestial but sometimes epiphytic.

Notes
- An endemic species.
- The commonest of our filmy ferns, especially in wetter areas.

▲ *Hymenophyllum dilatatum*

▼ *Hymenophyllum revolutum*

31 *Hymenophyllum dilatatum*
Filmy fern / Matua

Family HYMENOPHYLLACEAE

Description
- Rhizomes long-creeping and wiry.
- Fronds pale-green and delicate, with broad, smooth-margined pinnae segments.
- Sori conspicuous and broad.

Distribution & Habitat
- Common throughout country, except for drier parts of Canterbury and Otago.
- Usually epiphytic; sometimes grows on logs or ground.

Notes
- The largest of our filmy ferns.
- Endemic.

32 *Hymenophyllum revolutum*
Filmy fern / Mauku

Family HYMENOPHYLLACEAE

Description
- Rhizomes long-creeping, thin and wiry.
- Fronds small, darkish green, variable, with coarsely toothed margins.
- Sori large on short lateral segments.

Distribution & Habitat
- Common in damper lowland to moderate altitude forest from about Bay of Islands southwards.
- Usually grows on logs and rocks, but sometimes epiphytic.

Notes
- One of our smallest and commonest filmy ferns.

▲ *Hymenophyllum sanguinolentum*

▼ *Hypolepis rufobarbata*

33 *Hymenophyllum sanguinolentum*
Filmy fern / Piripiri

Family HYMENOPHYLLACEAE

Description
- Rhizomes long and wiry.
- Fronds olive-green and lacy, with smooth-margined pinnae, usually compact but sometimes with elongated tips.

Distribution & Habitat
- Common throughout country, growing both on ground and epiphytically.

Notes
- An endemic species.
- Smells like blood if crushed.
- The most drought-resistent of our filmy ferns, curling into a tight ball in dry weather.

34 *Hypolepis rufobarbata*
Ground fern / Pikopiko

Family DENNSTAEDTIACEAE

Description
- Rhizomes long-creeping and hairy with stipes arising at irregular intervals.
- Fronds reddish-brown and somewhat sticky to touch, arising at intervals on hairy, purplish-brown stipes from rhizome.

Distribution & Habitat
- Rare north of Auckland city, but common south to Stewart Island, except in south Canterbury and parts of Otago.
- Prefers wetter, more open areas.

Notes
- An endemic species.
- Easily cultivated but can become weedy.
- Insects often become stuck to hairs on stipes.

▲ *Lastreopsis glabella*

▼ *Lastreopsis hispida*

35 *Lastreopsis glabella*
Shield fern / Tuakura

Family DRYOPTERIDACEAE

Description
- Rhizomes erect and non-creeping.
- Fronds finely divided, bluish-green, slightly glossy above and soft to touch with smooth stipes.
- Sori widely scattered, giving this fern a speckled appearance.

Distribution & Habitat
- Common throughout the country.
- Grows on open lowland and floors of coastal forest, favouring rocky ground and stream banks.

Notes
- Plants contain a natural insecticide.

36 *Lastreopsis hispida*
Hairy fern / Tuakura

Family DRYOPTERIDACEAE

Description
- Rhizomes thick and long-creeping, with fronds arising at intervals.
- Young fronds are many hues of green, giving this fern a distinctive appearance.
- Mature fronds dark green with dryish texture.

Distribution & Habitat
- Sometimes epiphytic, *L. hispida* is one of the few ground ferns that can adapt to the dry conditions typical of tree-fern forests.
- Common from North Cape to the Chatham Islands, as well as in eastern Australia.

Notes
- Can be confused with *L. glabella*, but can be separated from this species by its hairy stipes.

Leptolepia novae-zelandiae
Leptolepia / Pikopiko

Family DENNSTAEDTIACEAE

Description
- Rhizomes creeping and many-branched, bearing fronds at irregular intervals.
- Fronds dark matt green, finely divided, giving this fern a distinctive appearance.
- Dryish to touch.

Distribution & Habitat
- Widely distributed to moderate altitudes throughout the country from Northland to Stewart Island and the Chathams, but nowhere abundant.
- A terrestrial fern that grows in sheltered areas of full shade.

◀ *Leptolepia novae-zelandiae*

▲ *Leptopteris hymenophylloides*

▼ *Leptopteris superba*

38 *Leptopteris hymenophylloides*
Single crepe fern / Heruheru

Family OSMUNDACEAE

Description
- Rhizomes erect, forming a stout, woody trunk in mature specimens.
- Fronds dark, oily, translucent green, very finely divided.

Distribution & Habitat
- Found throughout the three main islands and also on the Chatham Islands.
- In contrast to the closely related *L. superba*, this fern prefers sites higher up on valley sides or even on ridge tops.

Notes
- An endemic species.
- Can be confused with *L. superba*, which is a larger fern with more finely cut fronds.

39 *Leptopteris superba*
Prince of Wales feathers / Heruheru

Family OSMUNDACEAE

Description
- Rhizomes erect, often forming short, thick trunks in mature specimens.
- Fronds delicate, very finely cut, dark green and translucent.
- Sporangia not in discrete sori, but scattered on undersides of fronds.

Distribution & Habitat
- Relatively common in moist, sheltered forest from about Mt Pirongia southwards.
- A montane fern in the north, but found at lower altitudes in the south and on Stewart Island.

Notes
- An endemic species.
- An extremely beautiful fern, but very difficult to grow in cultivation.

▲ *Lindsaea trichomanoides*

▼ *Lygodium articulatum*

40 Lindsaea trichomanoides
Lindsaea

Family DENNSTAEDTIACEAE

Description
- Small, compact tufts arising at intervals from a long-creeping rhizome.
- Fronds bright green and delicate, drooping with smooth or toothed margins.

Distribution & Habitat
- Throughout the North Island and the Marlborough Sounds, as well as Westland and Stewart Island.
- Also found in Australia.
- Favours lowland forest, preferring areas with better drainage.

Notes
- Carcinogenic.

41 Lygodium articulatum
Mangemange

Family SCHIZAEACEAE

Description
- Rhizomes creeping underground.
- Fertile and sterile pinnae remarkably different and both can occur on the same frond.
- Sterile pinnae oblong, yellow-green with smooth margins.
- Fertile pinnae are much smaller, multi-branched and deeply lobed.
- Sori are arranged in clusters on leaf margins.

Distribution & Habitat
- From North Cape south to about Rotorua.
- A climbing fern in lowland forest.

Notes
- The long, climbing fronds of *L. articulatum*, reaching up to 10 m in length, are the longest known of any plant.

Marattia salicina
King fern / Para

Family : MARATTIACEAE

Description
- Rhizomes erect and compact.
- Newly emergent fronds are translucent yellowish-green, darkening as fronds mature.
- Fronds grow up to 400 cm in length.

Distribution & Habitat
- In lowland forest north of a line from the Bay of Plenty to Taranaki.
- Also found in Queensland, Australia and on Norfolk Island and Lord Howe Island.

Notes
- This is by far the largest of our ground ferns.
- Once an important food plant of the Maori, King fern has now been wiped out by pigs in many areas.

◀ *Marattia salicina*

▲ *Microsorum pustulatus*

▼ *Microsorum scandens*

43 *Microsorum pustulatus*
Hound's tongue / Kowaowao

Family POLYPODIACEAE

Description
- Rhizomes stout and creeping.
- Fronds glossy green and thick.
- Fronds can be a variety of forms, from strap-shaped to lobed.
- Sori are prominent and brown, causing raised bumps on upper surface of frond.

Distribution & Habitat
- This fern is found on ground and rocks.
- Also epiphytic in scrub and forest from coastal to subalpine areas.

- Common throughout the country from the Kermadecs to the sub-Antarctic islands.
- Also on Norfolk Island, Tasmania and in southeast Australia.

Notes
- Previously known as *Phymatosorus diversifolius*.
- In Australia it is called the Kangaroo fern.
- Maori ate the emergent fronds of this fern.

44 *Microsorum scandens*
Fragrant fern / Mokimoki

Family POLYPODIACEAE

Description
- Rhizomes slender and creeping.
- Fronds very thin in texture, dark green and often pendulous.
- Sori on lower surface prominent and dark brown, causing bumps on upper surface of frond.

Distribution & Habitat
- Throughout North Island and in the north and west of the South Island.

- Also found in Australia.
- Epiphytic on tree trunks; sometimes grows on rocks and rocky banks.

Notes
- Previously known as *Phymatosorus scandens*.
- The common name refers to the faint but distinctive odour of the living plant, which has been likened to marzipan.

▲ *Nephrolepis* sp.

▼ *Paesia scaberula*

45 *Nephrolepis* sp.
Thermal sword fern

Family DAVALLIACEAE

Description
- Rhizomes short and erect.
- Fronds yellowish-green, divided into numerous sickle-shaped or linear segments.
- Sori arranged into two rows of kidney-shaped clusters close to the margins of the leaf segments.

Distribution & Habitat
- In open forest on the Kermadec Islands, otherwise confined to the thermal areas of the central North Island.

Notes
- As yet has no scientific name.
- Very similar to the introduced *N. cordifolia*, but it differs from this fern in having no tubers on its runners.

46 *Paesia scaberula*
Ring fern / Matata

Family DENNSTAEDTIACEAE

Description
- Rhizomes creeping and wiry.
- Fronds very finely divided, yellowish-green, harsh to touch, somewhat sticky when young.
- Strongly scented where it grows in open areas.

Distribution & Habitat
- Forms large patches in suitable areas such as disturbed ground, roadside cuttings and along margins of regenerating bush.

Notes
- This fern is poisonous to stock and is considered to be a weed.

▲ *Pellaea rotundifolia*

▼ *Pneumatopteris pennigera*

47 *Pellaea rotundifolia*
Button fern / Tarawera

Family PTERIDACEAE

Description
- Rhizomes much branched and wiry with fronds arising at irregular intervals.
- Fronds deep green above and paler below, with rounded or oval pinnae.

Distribution & Habitat
- Common throughout the country, from the Three Kings Islands to Otago.
- Occurs in rocky areas, open forest and light scrub.

Notes
- An endemic species, but now available commercially in many countries under the name Button fern.

48 *Pneumatopteris pennigera*
Gully fern / Pakauroharoha

Family THELYPTERIDACEAE

Description
- Rhizomes short and erect, developing into a short trunk in mature specimens.
- Fronds light green and deeply lobed with golden-brown stipes arising in a distinctive symmetrical pattern.

Distribution & Habitat
- Common in the North and South Islands and on the Chathams, but absent from Stewart Island.
- Also found in eastern Australia, where it is rare and local.
- Grows on stream banks and in damp gullies in deep shade.

Notes
- Edible.
- In Australia this species is known as Lime fern.

▲ *Polystichum richardii*

▼ *Polystichum vestitum*

49 *Polystichum richardii*
Common shield fern / Pikopiko
Family DRYOPTERIDACEAE

Description
- Newly emergent fronds pale green and soft, becoming bluish-green with distinctive dark midribs in mature specimens. Harsh to touch.
- In Wellington area this fern is noticeably darker.

Distribution & Habitat
- Common in coastal areas of the North Island and in eastern areas of the South Island.
- On coastal rocks, along bush margins, in open scrub and forest, and on dry banks, sometimes under pohutukawa.

Notes
- An endemic species.
- Emergent fronds are edible but should only be used in an emergency.

50 *Polystichum vestitum*
Prickly shield fern / Puniu
Family DRYOPTERIDACEAE

Description
- Rhizomes erect, forming a trunk in older specimens.
- Fronds glossy, dark green and prickly, arising in a distinctive crown.

Distribution & Habitat
- Common in the South Island and Stewart Island, but occurs more locally and at higher altitudes in the North Island.
- On open ground, among sub-alpine and alpine scrub and in forest, often persisting after bush has been cleared.

Notes
- An endemic species.
- An enthusiastic coloniser of open ground.
- Young plants are susceptible to deer and possum damage.

▲ *Pteridium esculentum*

▼*Pteris comans*

51 *Pteridium esculentum*
Bracken / Rahurahu

Family DENNSTAEDTIACEAE

Description
- Rhizomes thick, creeping, underground.
- Fronds dark green to brownish-green and wiry. Hairless when old.
- Sori in continuous line along pinnae margins

Distribution & Habitat
- Common in both lowland and montane areas, particularly along bush margins and on disturbed ground.

- Found throughout the country.
- Also occurs on offshore islands, in Australia and many Pacific Islands.

Notes
- Poisonous to stock.
- Edible rhizomes called 'aruhe' were a staple food of the Maori, but are carcinogenic in excess.
- Used by settlers as substitute for hops in brewing beer.

52 *Pteris comans*
Coastal brake

Family PTERIDACEAE

Description
- Rhizomes short and erect.
- Fronds pale green, broadly triangular and much divided.
- Sori in continuous lines of varying lengths along pinnae margins.

Distribution & Habitat
- Coastal areas from Northland to the Bay of Plenty; also on the Kermadec Islands, Three Kings Islands, and islands of the Hauraki Gulf.
- Also in eastern Australia.
- In coastal scrub and pohutukawa forest.

Notes
- Could be confused with other *Pteris* species, but *P. comans* has much broader pinnules.

▲ *Pteris tremula*

▼ *Pteris macilenta*

53 *Pteris tremula*
Shaking brake / Turawera

Family PTERIDACEAE

Description
- Rhizomes short and erect.
- Fronds pale green, arising in tufts.
- Sori in continuous lines of varying lengths along pinnae margins.

Distribution & Habitat
- Found from Northland to Otago, but absent from Stewart Island.

- Also in Australia.
- Favours drier areas and dies back in adverse conditions.

Notes
- The common name refers to the way the fronds quiver in any breeze.
- Smells like tom cat urine when broken.

54 *Pteris macilenta*
Sweet fern / Titipo

Family PTERIDACEAE

Description
- Rhizomes short and erect.
- Fronds light green, much dissected, thin and widely spaced.
- Sori in continuous lines of varying lengths along pinnae margins.

Distribution & Habitat
- Common in the North Island in coastal and lowland areas, as well as in northern and western

areas of the South Island. Also found on the Chatham Islands.
- Grows in drier, more open scrub and forest, and on coastal cliffs.

Notes
- The common name of this fern is a mystery, as it neither smells nor tastes sweet.

▲ *Pyrrosia eleagnifolia*

▼ *Sticherus cunninghamii*

55 *Pyrrosia eleagnifolia*
Leather-leaf fern / Ngarara wehi

Family POLYPODIACEAE

Description
- Sterile fronds are dark green, thick, leathery, short and broad, glossy on top, with numerous whitish hairs below.
- Fertile fronds are much longer, with conspicuous brown sori.

Distribution & Habitat
- Common in both lowland and montane areas throughout the country from the Kermadec Islands to the Chathams.
- Epiphytic on trees and sometimes on fallen logs.
- Grows on rocks and over stone walls in Auckland and Northland.

Notes
- Can be grown in rockeries, but difficult to establish.

56 *Sticherus cunninghamii*
Umbrella fern / Tapuwae kotuku

Family GLEICHENIACEAE

Description
- Rhizomes creeping and much branched.
- Fronds bright green, much paler below and divided into umbrella-shaped tiers.

Distribution & Habitat
- At higher altitudes in the North Island, descending to coastal areas in the South Island and Stewart Island.

Notes
- An endemic species.
- The Maori name means 'footprint of the white heron', reflecting its distinctive shape.

▲ *Sticherus flabellatus*

▼ *Trichomanes reniforme*

57 *Sticherus flabellatus*
Shiny fan fern / Waekura

Family GLEICHENIACEAE

Description
- Rhizomes wide-creeping.
- Fronds erect and spreading in umbrella-like tiers, shiny dark green above, but paler below.

Distribution & Habitat
- Somewhat disjunctive in distribution. Common in the gumlands of the far North; locally plentiful in areas south of Auckland and in the northern and western areas of the South Island.
- Also found in Tasmania, where it is regarded as a rare fern, as well as in Papua New Guinea and New Caledonia.

Notes
- Could be confused with *S. cunninghamii*, but this is larger plant with narrower, darker green fronds.

58 *Trichomanes reniforme*
Kidney fern / Raurenga

Family HYMENOPHYLLACEAE

Description
- Rhizomes long-creeping.
- Fronds distinctively bright, glossy, pellucid green, round or kidney-shaped.
- Sori crowded on leaf margins.

Distribution & Habitat
- Found throughout the three main islands and on the Chatham Islands.
- Occurs locally, but is common in areas where it does occur. Sometimes forms extensive mats.
- Grows on rocks and as an epiphyte on trees.

Notes
- An endemic species.
- Shrivels up in dry weather, expanding again as soon as rain falls.
- Traditionally used by Maori in the treatment of bowel disorders.

▲ *Lycopodium cernuum*   ▼ *Lycopodium deuterodensum*

Native fern allies

59 *Lycopodium cernuum*
Creeping club-moss / Matukutuku

Family LYCOPODIACEAE

Description
- The main stem is looping with roots at intervals and aerial stems arising midway between rooting points, making it resemble a young fir tree.
- Yellowish-green in colour.
- Stems covered with small, awl-shaped leaves.

Distribution & Habitat
- From the Kermadec Islands to northern Westland.
- Also widespread throughout the tropics and subtropics.
- Occurs in scrub, along bush margins, roadside cuttings and some thermal areas.

Notes
- Traditionally used by Maori to treat chest complaints and rheumatism.

60 *Lycopodium deuterodensum*
Tree club-moss / Puakarimu

Family LYCOPODIACEAE

Description
- Erect, much-branched aerial stems arising from creeping rhizomes resemble young rimu trees.
- Ranges in colour from dull green to orange, depending on habitat.
- Stems covered with small, scale-like leaves, or can have longer, fluffy leaves.

Distribution & Habitat
- From the Kermadec Islands to northern Westland.
- Also found in New Caledonia.

Notes
- The Maori name refers to the plant's similarity to a rimu.
- The scientific name refers to this plant's two distinctly different foliage forms.

Lycopodium volubile
Climbing club-moss / Waewaekoukou

Family LYCOPODIACEAE

Description
- Branched stems arising from creeping rhizomes crawl over the ground and twine up and through vegetation.
- Ranges in colour from dull green to yellow, depending on habitat.
- Stems covered with small, scale-like leaves.

Distribution & Habitat
- From the Kermadec Islands to about Banks Peninsula.

- Occurs from coastal to montane areas on disturbed ground such as roadside cuttings and along bush margins.
- Also found in Asia, Papua New Guinea and New Caledonia.

Notes
- Used to make head-bands by Maori women in times of mourning.

◀ *Lycopodium volubile*

Tmesipteris elongata
Fork fern

Family PSILOTACEAE

Description
- Rhizomes give rise directly to pendulous unbranched or simply-divided stems with dull-green leaves.
- Spore cases are borne in notches at the bases of these leaves.

Distribution & Habitat
- Epiphytic, mostly on tree ferns, in the North Island, the north and west of the South Island, and on Stewart Island and in the Chathams.
- Also occurs in Victoria and Tasmania.

Notes
- The most ancient member of allied fern family. Plants with similar stem structures grew in the Devonian period 400 million years ago.

◀ *Tmesipteris elongata*

▲ *Nephrolepis cordifolia*   ▼ *Pteris cretica*

Exotic ferns and fern allies

63 *Nephrolepis cordifolia*
Boston fern

Family DAVALLIACEAE

Description
- Rhizomes short and erect with fleshy tubers.
- Fronds yellowish-green, erect with over 50 pairs of pinnae.
- Sori in raised kidney-shaped clusters near leaf margins.

Distribution & Habitat
- Widespread on disturbed ground and in gardens from Northland to Canterbury.
- Sometimes epiphytic; also grows over rocks.

Notes
- Tubers are edible.
- A common fern in cultivation, which assists in its dissemination.

64 *Pteris cretica*
Cretan brake

Family PTERIDACEAE

Description
- Rhizomes short-creeping
- Fronds long, ovate, with smooth margins, dull-green in colour.

Distribution & Habitat
- A naturalised species that is now wide-spread and locally common from Northland to Banks Peninsula.
- Also occurs in southern Europe, Africa and Asia.
- Prefers disturbed ground and cemeteries.

Notes
- Can be readily distinguished from native species by the long, smooth-edged leaf margins.
- Many different forms have been spread by gardeners.

▲ *Selaginella kraussiana*

Selaginella kraussiana
Selaginella

Family SELAGINELLACEAE

Description
- Stems creeping, randomly branched, forming large mats.
- Sporangia arranged in cones in axils of leaves (i.e., the point where the leaf joins the stem).

Distribution & Habitat
- A naturalised species found in many lowland areas of the North Island as well as in wetter areas of the South Island.
- Terrestrial.

Notes
- A serious nuisance in many wooded areas, where it is a vigorous coloniser, crowding out native plants.
- Originally from South Africa.

Further Reading

Brownsey, Patrick J. and Smith-Dodsworth, John C.
New Zealand Ferns and Allied Plants.
David Bateman Ltd, Auckland, 1989.

Chinnock, R.J. and Heath, Eric.
The Reed Handbook of Common New Zealand Ferns and Fern Allies.
New Edition. Reed Books, Auckland, 1999.

Van der Mast, S. and Hobbs, J.
Ferns for New Zealand Gardens.
Godwit Publishing, Auckland, 1998.

Index of Common and Maori Names

Index of Scientific Names